ISBN – 10: 1456418238

First Published November 2010

This Edition – February 2011

The

Message

For my daughter Erin, always believe in yourself and never stop searching for answers.....

PART I

THE RIDDLE

CHAPTER 1

Mount Olympus

We are made in the image of the creator.

What does this mean? Does it mean we look like God? I think it simply means we are made from the essence or mind of whatever made us. This makes sense. If you are an artist and you paint something or write something, that product is made from your WILL. It is a reflection of you. When I refer to God or the Creator in this book I am simply referring to whatever force created the universe and not to any single or specific deity.

To start, let us assume we are made in the image of God. All our traits or characteristics must therefore be Godlike or of God. This would imply that all the good and bad inside us must be a reflection of the true nature of our creator. Even if you do not believe in God, if you simply believe that we are here thanks solely to gravity, then we are still only a product of the forces that made us. Where does this leave us?

As humans we are always creating things ourselves. People like to invent and build things. Right now we are making our computers smarter and smarter, hoping that one day we will be able to create artificial intelligence. This will be a computer that can think and reason for itself. Recently IBM released a supercomputer named Watson that is able to compete with its human opponents on the hit television game show Jeopardy. Although it is not considered artificial intelligence yet, it is a big step in this direction. Watson is able to analyze incredible amounts of data and uses state of the art voice recognition software to interact with its human counterparts; Watson however is incapable of making simple judgement calls. While programmers are working diligently toward this goal, doctors, biologists and geneticists are trying to create artificial life, these are living organisms trying to create other living organisms for themselves. We want the power that God has, we are made in the image of God and we want to be like God, no matter what God is.

Now, let's hit fast forward and move a billion years into the future. Based on our current rate of technology, what will this future be like? For anyone that is not up to speed with exactly where the human race is in terms of technology, let me give you a crash course. Right now, in 2010 we can accelerate sub-atomic particles to almost the speed of light, we can stop the human heart using a heart-lung machine while we drain all the blood out of the body in order to perform open heart surgery, we are thinking about or maybe even actively experimenting on creating nanobots that can capture every thought inside our heads and transfer those thoughts into a computer in order to preserve our "self". The Human Genome Project (HGP) has already successfully identified all the genes in human DNA and now continues to analyze this data.

Where will we be in a billion years? Will we be able to create artificial intelligence? Will we be able to create life from scratch? Will we be able to create an independent environment or virtual world to house this new life? Still operating under the assumption that we are made in the image of God, what would we do with all this accumulated power?

Humans are creative; we also like sex, violence, art, music, mystery, suspense and drama. We are arrogant, funny, selfish, moody, loving, kind, and all the other things that make us so diverse and so unique. So what would we do with all the power of God? I think the answer to this and to every other question that haunts humanity can be seen reflected back at us from our own society and our creations, we can see it in the movies we produce, in our art, in our stories we write, in our games we program and in almost every aspect of our creation.

Harnessing the power of the Gods we could put on the greatest show on earth, actually the greatest show in the universe. With the power of God, we could create our own virtual or replica universe and fill it with intelligent, living creatures that we create. This would be zillions of times better and more realistic than any movie or computer game we have ever made. But how could we make it really entertaining? We don't want to see the same boring old movies and plays we are accustomed to. For Pete's sake we don't want to see a rewrite of Dallas. We want something fresh and exciting, something unpredictable. You know what we could do? We could give our intelligent creatures their own free will. Instead of having them live and act according to a tired old script, we could let them reason and decide what to do for themselves. This will be the most capricious extravaganza ever released. We could play God.

Okay, so now the stage is set. We have created our Universe and sheathed ourselves in mystery. We have left the entire production open for debate, in order to really confuse our leading characters we allow them to evolve from simple mindless beasts into great thinkers, all this from a puddle of primordial soup, this way there is no trail leading directly back to us. What could be better? Well, there is one thing that could be better but I'll tell you later. Right now, all that's left to do is to yell lights, camera and action, sit back and enjoy the show.

Chapter 2

Let there be light

The event starts with a dramatic flash, not really a bang because there is no sound before the Universe or any light for that matter. It simply starts....

According to scientific evidence, the Universe as we (humans) know it came into existence about 13.7 Billion years ago after a singularity event known as the big bang. Renowned Physicist Stephen Hawking theorises that this singularity event could have been brought about entirely by the force of gravity, without any need for divine intervention. Whether or not this is accurate, or if in fact God (some super powerful and super intelligent entity or entities) created the Universe, the evidence seems to suggest that whatever happened happened about 13.7 billion years ago. The truth may lie behind an impenetrable curtain of cosmic microwave background radiation.

As humans the first thing to note is that 13.7 billion years is simply our way of calculating time. There is no universal standard of time. Time only came into existence after the universe was born and time as we know it only exists inside our universe. When physicists view time from different points in space or at different velocities, time appears to differ for each person depending on their position and speed relative to each observer, hence the term "time is relative to the observer". This observation is crucial to our perception and understanding of creation.

In the early 1900's it was the great physicist Albert Einstein who gave us the formula $E=MC^2$ along with his theory of special relativity, which despite its familiarity is something most people do not understand. Einstein realized that matter (mass) and energy are essentially different forms of the same thing and that one could be converted into the other and vice versa. This conversion only required motion such as kinetic energy, heat, or light. Einstein further theorized that nothing will ever be able to move faster than light. In the equation $E=MC^2$, the E represents energy, M represents mass and C^2 is the square of the speed of light.

Thanks to Einstein's theory of special relativity we know that if you begin accelerating through space at phenomenal speeds, for you time slows down and has less effect on you while the rest of the universe ages at a normal pace. This phenomenon is known as time dilation. The only thing we really need to know right now is that time is relative to the observer and that two people experiencing the same event from different points in space or moving at different velocities cannot agree on when exactly that event took place.

We will come back to this topic later as I believe a large answer to the mysteries behind the creation of the Universe lies hidden in Einstein's work.

Immediately after the big bang or creation event the universe was a very violent and volatile environment to be in, from what we can see as we look back over the past, we can conclude that the universe was initially a very small and tightly condensed place or sphere that has been expanding for the last 13.7 billion years. Physicists and cosmologists are not entirely sure what force is driving this expansion; they do not know for how much longer we will keep expanding or what will happen when this expansion stops. Will the universe tear itself apart in a big rip? Will it start shrinking into a tightly compressed ball once again or will it reach a steady or frozen state like a clock that simply runs down and stays frozen for the rest of eternity? I think part of the uncertainty or confusion comes from thinking of the universe as a place or thing at all.

Whatever the ultimate fate of the universe, I believe the answer to this question is something that will greatly affect us all, no matter how far away into the future this end may come about, even though anybody currently alive will be long dead by this time, I think the answer to what the fate of the universe is, is something very relevant to every creature that has ever lived or will ever live.

There are numerous and varying opinions on exactly what the universe is and on how or what created the universe. Amongst these, the leading two schools of thought are probably:

1. The Universe is a totally random sequence of events, created purely by nature and physics such as gravity.

2. The Universe is created by a vastly superior intellect or God if you will, for a specific purpose.

Of course there are other theories; some believe the universe is random but that God created it to be so. Others will tell you that God created the universe but has since abandoned the project and forgotten about us. The themes are numerous and varying, yet somebody must be right.

CHAPTER 3

Religion

The Oxford dictionary definition of religion is:

> Belief in super-human controlling power, esp. in personal God or gods entitled to obedience; system of faith and worship.

There are numerous religious groups around the world today and more seem to spring up as time goes by. Those with the greatest worldwide following are probably Christianity, Islam, Hinduism, Buddhism and Judaism. But even these can be broken up into smaller sects as it is very hard to get everyone to agree on precisely what happened before the dawn of time and what exactly our reason is for being here.

The mere fact that religions around the world have a combined support of more than 80% of the world's population would suggest that people are more inclined to believe that some omnipotent deity created the universe for a higher purpose. This does not necessarily make it true. But whether or not it is a question of natural instinct, pure faith or mere ignorance, we cannot simply disregard the opinion of billions of people. If we are indeed the product of whatever created us, the majority of us believe that we were created by a God of some sort in order to fulfill a specific purpose.

The first demographic to consider is that most of these religious followers (though certainly not all) are unsophisticated, poor people living according to customs bestowed upon them by previous generations. However, the fact that they are not formally educated does not mean that they are unintelligent or incapable of drawing logical conclusions about the nature of the universe. Logic and reason are powerful and natural tools that do not require formal education. And maybe people who are more in touch with humanity and suffering on a daily basis are more exposed to the hand of God than people seeking instant gratification, surrounded by shiny objects, modern technology and material wealth.

Whatever the reasons for their belief, in order to reach true enlightenment one must be prepared to consider the opinions of all regardless of their station in life or modern society.

Christianity

Christianity is a monotheistic religion based on the Old Testament and the teachings of Jesus as embodied in the New Testament. Followers of Christianity believe that the Bible is the written word of God and that Jesus was literally the son of God.

Christianity teaches that there is only one God and that God made the universe, the Earth, and created Adam and Eve all over a period of six days. God created man in His image. This does not mean that God has a human body or form. Image refers to the likeness of God's character and rationality. We are a reflection of God. This also implies that we did not evolve through a process of random selection from single-celled organism into rational, emotional beings but that we were created fully formed by the hand of God.

Christians believe that mankind was created specifically to have a relationship with God, but sin separates all men from God. In order to be saved and be granted entrance into heaven after death, one must place one's faith entirely in the finished work of Christ on the cross.

Some Christians firmly believe that the entire Universe is only about 6000 years old and that the end of days is close at hand. This end of days will be preceded by an event called the rapture. During the rapture Gods chosen will be snatched out of this universe and spared the horrors that will come about during Armageddon. The dead will rise again and every person who has ever lived will be judged by God at this end of time.

Islam

One of the most distinguishing features of the Islamic religion is their deep faith in God (Allah) and their conviction that whatever happens in the universe and whatever befalls them, only happens through the will and the decree of God. Muslims believe that God is one and incomparable and that the purpose of life is to worship God.

Islam means acceptance of and submission to God and believers must demonstrate this by worshiping Allah, following his commands, and avoiding polytheism.

This belief and worship are based upon the teachings of the Qur'an and the authentic Hadith. A Muslim feels that he is in constant need of the help and support of God. He also has no choice in his life but to submit to the will of God, worship Him, strive towards the Right Path and do good deeds. This type of mentality will guide him to be righteous and upright in all his deeds, both in public and in private.

Islam does not have a set position on the actual age of the Universe, leaving such knowledge to God alone. As with Christianity, the Qur'an also describes the creation of the universe in "six days," but then in other passages indicates that a "day" to us is different than a "day" in God's reckoning of time.

Hinduism

Hinduism is the world's third largest religion, after Christianity and Islam, with more than a billion followers. Hinduism is formed of diverse traditions and has no single founder.

The Hindu faith has many gods and goddesses and is therefore polytheistic. According to Hinduism, three Lords rule the world. Brahma: the creator; Vishnu: the preserver and Shiva: the destroyer.

Though Hindus worship different idols or gods, many Hindus do believe in one God and view all these different gods and goddesses as different faces or aspects of the same true God.

Hindu's do not have a Bible or Quran but there is a large body of texts that are divided into Śruti (revealed) and Smriti (remembered) texts. These texts discuss theology, philosophy and mythology, and provide information on the practice of dharma or religious living.

Hindus believe in reincarnation. Reincarnation is the rebirth of a soul in a continuous cycle of life and death. The basic belief is that a person's fate is determined according to his deeds. These deeds in Hinduism are called 'Karma'. A soul who does good Karma in this life will be awarded with a better life in the next incarnation. Souls who do bad Karma will be punished for their sins, if not in this incarnation then in the next incarnation and will continue to be born in this world again and again. The good souls will be liberated from the circle of rebirth and get redemption which is called 'Moksha' meaning freedom.

According to Hindu Philosophy, the universe (or multiverse) never came to be at some particular point, but always has been, always will be, and is perpetually in flux. Space and time are of cyclical nature. This universe is simply the current one, which is in flux and constantly changing, when it finally ceases to manifest, a new one will arise.

There are many metaphorical parallels between modern science and the Hindu theory of 'Shrishti'. The 'Anda' itself resembles the hypothetical energy point from which the Big Bang and hence the Universe was created according to modern science. The true intonation of Om is very long and drawn out, it is described as an all pervading sound. Its parallel in modern cosmology is the cosmic background radiation which pervades the entire Universe.

Even the end of the universe as described by some physicists favouring a big crunch holds a great parallel to the Hindu idea of an expanding universe which then contracts in a big crunch from which a new universe is born.

Hinduism is the only religion in which the time scales correspond to those of modern scientific cosmology believing that the universe is billions of years old.

Buddhism

Buddhism is a religion and philosophy revolving around a variety of traditions, beliefs and practices, largely based on teachings attributed to the "Buddha".

The Buddha is recognized by Buddhists as an awakened or enlightened teacher who shared his insights to help sentient beings end suffering, achieve nirvana, and escape the otherwise endless cycle of suffering and rebirth.

In Buddhism, karma specifically refers to a person's actions (of body, speech, and mind) that spring from mental intent and which bring about a consequence or result. Every time a person acts there is some quality of intention at the base of the mind and it is that quality rather than the outward appearance of the action that determines its effect.

Rebirth in Buddhism refers to a process whereby beings go through a succession of lifetimes as one of many possible forms of sentient life, each running from conception to death. Buddhism rejects the idea of a permanent self or an unchanging, eternal soul, as it is defined in other popular religions.

According to Buddhism there is ultimately no such thing as a self independent from the rest of the universe. Rebirth in this and subsequent existences must be seen as the continuation of a dynamic, ever-changing process of "dependent arising" determined by the laws of cause and effect (karma) rather than that of one being, transmigrating or incarnating from one existence to the next.

Judaism

Judaism is one of the oldest monotheistic religions and dates back to over 3000 years. Judaism is a way of life that is considered by the Jews to be an expression of their covenantal relationship with God.

Orthodox Jews believe that the Torah sometimes known as the Pentateuch (First 5 books of the Bible) and Jewish law are divine in origin, eternal and unalterable, and that they should be strictly followed. They believe that these laws were passed down directly from God to Moses on Mount Sinai

Jews believe that God is directly concerned with the actions of mankind. God commands all Jewish people to imitate his love for them and to reciprocate this love by caring for and loving each other.

According to Jewish cosmology God (Yahweh) created all things through a series of acts, or methodology, as described in the Bible.

In the beginning God created the heaven and earth. And the earth was without form, and void; and darkness was upon the deep. And the spirit of God moved upon the face of the waters. In the first day God said "let there be light", created light, and there was light; God saw the light, and it was good. God moved the light from darkness; calling the light Day and the darkness Night

God made man in his own image, both male and female. And God blessed them and instructed them as to how he wanted them to live upon the earth.

God is the source of everything that exists.

Interestingly enough the Bible Code, a best-selling book written by Michael Drosnin claims to be a mathematically proven theory that the first 5 books of the Bible (The Torah) which was given directly to Moses has been coded with secret texts that can be used to predict the future.

It is thought that this coding was done by highly advanced beings or even by God and can only be decrypted using modern day computers.

Even the great scientist Isaac Newton was obsessed with finding a hidden code inside the Bible (Torah).

As you can tell from these predominant religions, the nature of existence and the reason for being varies greatly. So what is the truth? Can they all be right? I think in some small way they all contribute to the true nature of existence. I believe these religions are so popular because they all hold some essence of the truth inside them. I also believe that the answers to life are all around us and that is why we latch onto certain religions because in some small way they all capture some part of what we intrinsically feel deep down inside us.

There are of course many other religions and it is always fascinating to read up on what people believe, I urge you to do so and to always keep an open mind.

CHAPTER 4

Ancient Cultures and beliefs

Aside from popular modern religion and their take on creation there were also great ancient cultures that adopted their own perspectives on the universe.

The Maya

The Maya are probably the best-known of the Mesoamerican civilizations. Mayan history dates back to around 2600 B.C.

The Maya believed in a cyclical nature of time. Their rituals and ceremonies were very closely associated with celestial and terrestrial cycles which they observed and inscribed as separate calendars.

The Mayans believed that the Cosmos could be divided into three parts, the Earth, the underworld beneath and the heavens above. They worshipped everything in nature and tried to explain how things happened because of the Gods.

The Mayans often made sacrifices to the Gods. In some Maya rituals people were killed by having their arms and legs held while a priest cut the person's chest open and tore out his heart as an offering to the Gods.

The natural cycles of observable phenomena, and the recurrence and renewal of death-rebirth imagery in their mythological traditions were important influences upon Maya societies.

According to Maya cosmology the present world and the humans in it were preceded by other worlds (one to five others, depending on the tradition) which were fashioned in various forms by the gods, but subsequently destroyed. The present world also had a tenuous existence, requiring the supplication and offerings of periodic sacrifice to maintain the balance of continuing existence.

Although many people today believe that in accordance with Mayan tradition and the Mayan calendar the world is destined to end on 21st of December 2012, there are many others who believe that this simply marks a time of transition from one World Age into another. This will be a time of great revelation when a veil will be lifted on secret knowledge that has to date been hidden from mankind.

The Aztecs

The Aztec world also consisted of three main parts, the earth world on which humans lived, an underworld which belonged to the dead and the upper plane in the sky. The earth and the nether world were both open for humans to enter, whereas the upper plane in the sky was impenetrable to humans.

Existence was envisioned as straddling the two worlds in a cycle of birth, life, death and rebirth.

The Aztecs practiced human sacrifice on a monumental scale. They also share the Mayan believe that the universe exists as great waves of energy that recycle over time. As a wave ripples through creation its movement synchronizes nature, life and time with its passing.

According to the Aztec calendar our current period in time marks the beginning of a new cycle.

The Egyptians

Egyptian religion was based on their interaction with a multitude of Gods who were believed to be present in, and in control of, the forces and elements of nature. The myths about these gods were meant to explain the origins and behavior of the forces they represented, and the practices of Egyptian religion were efforts to provide for the gods and gain their favour.

Religious practice centred on the pharaoh, the king of Egypt. By virtue of his kingship the pharaoh was believed to possess a divine power. He acted as the intermediary between his people and the gods, and was obligated to sustain the gods through rituals and offerings so that they could maintain order in the universe.

The Egyptian conception of the universe centered on maat, a word that encompasses several concepts in English, including "truth," "justice," and "order." Maat was the fixed, eternal order of the universe, both in the cosmos and in human society. It had existed since the creation of the world, and without it the world would lose its cohesion. In Egyptian belief, maat was constantly under threat from the forces of disorder, so all of society was required to maintain it. On the human level this meant that all members of society should cooperate and coexist; on the cosmic level it meant that all of the forces of nature (the gods) should continue to function in balance.

The most important part of the Egyptian view of the cosmos was the conception of time, which was greatly concerned with the maintenance of maat. Throughout the linear passage of time, a cyclical pattern recurred, in which maat was renewed by periodic events which echoed the original creation.

Overall Egyptians appeared to be more concerned with their continued existence in this world rather than a union with the Gods after death.

The Ancient Greeks

Ancient Greek religion was polytheistic, consisting of the worship of many Gods. Their belief's were identical to other earlier forms of Paganism. There was a hierarchy of Gods, with Zeus, the king of the gods, having a level of control over all the others.

The Greeks thought that the Gods would offer protection and guide their city-states. In many ways the gods behaved like human beings, and had human vices. They would often interact with humans, sometimes even spawning children with them.

Religion was a central part of Greek society and culture; they performed sacrifices and regularly worshiped the Gods in order to unify the people and to appease the Gods. The Greeks also believed that an afterlife awaited them after their death.

The afterlife was very important to the Greeks, as they believed that their souls were carried on to another dimension and lived on after death. This belief in an afterlife was a direct contrast to other forms of Paganism as most Pagan religions believed that there is no Heaven or Hell, Pagans tend to believe in re-incarnation and that the soul gets passed on to another body after death.

There was no singular Greek cosmogony, or creation myth. Different religious groups believed that the world had been created in different ways.

The Romans

Early Roman religion revolved around spirits. These spirits were thought to influence daily life for either good or evil purposes. The Romans therefore had to keep the spirits happy through constant worship and sacrifice in order to gain their favour.

Each of these spirits or Gods had a specific sphere of influence and the rituals to please these spirits had grown very complex over time. The Romans later appointed Priests to carry out these sacred rites.

The Ancient Romans, despite funeral rituals and annual celebrations of the dead, left no detailed literature of an afterlife and may have paid little attention to such concepts.
Roman religion was transformed by contact with conquered regions, introducing new gods and goddesses as well as new forms of worship often associated with mystery cults.

These are only some of the ancient religions and practices of the people who ruled the earth long ago and helped to shape our modern day society and religions.

The predominant or recurring themes for me is the belief in a force that created the universe, the idea that this force may in fact interact with humans or watch over us and also the idea that this universe is somehow eternal and our lives are not fixed but somehow in flux and we are continuously reincarnated either in this same universe or in alternate universes which some call heaven or nirvana.

CHAPTER 5

Philosophy

Religion or a belief in the super-natural was not the only way in which people have tried to explain and comprehend the universe. Apart from the spiritualists there were also those who considered themselves to be the rationalists. Many of these philosophers have greatly influenced our current understanding of the world and I personally believe that in order to completely appreciate the cosmos the person on the path to enlightenment needs to embrace science, spirituality and philosophy in equal measures.

Plato

In Plato's theory of the 'Forms' he asserts that abstract ideas possess the highest and most fundamental kind of reality and not the material world of change as we experience it through our senses. He believed that it was only through studying these Forms that we could gain any real knowledge about life.

According to Plato the objects that we see in this world (including ourselves) are not real but are only copies or mere shadows of the true Forms that are housed somewhere else. Plato describes the world of Forms as a pristine region of the physical universe located in a "place beyond heaven".

These Forms are the essences of everything we experience, they are that without which a thing would not be the kind of thing it is, examples are humans, animals, trees, mountains, love, beauty, colors and all the things that form the basis of our reality.

Plato's Forms do not exist at any specific point in time or space and have no spatial dimensions whatsoever, they are neither eternal in the sense of existing forever or mortal, of limited duration, they exist totally transcendent to time and space altogether.

Forms are totally non-physical, but they are not in the mind. They are extra-mental and considered to be REAL in the strictest sense of the word and are the basis or "blueprint" of perfection and remain perfect and unchanging.

For example, if we see an apple, that apple is only a shadow of the perfect Form that allows us to know that what we are seeing is an apple.

Aristotle

When it came to the question of the Universe Aristotle regarded the world as being uncreated and eternal. In Book Eight of his Physics, he describes what he calls the "Unmoved Mover" or "Prime Mover," which is the ultimate source, or cause, of motion in the universe, but is itself unmoved. For Aristotle this is God, who dwells at the circumference of the universe and causes the first movement in creation. The closer to the Unmoved Mover a body is, the more quickly it moves. Although the Unmoved Mover is God, it did not create the world.

Aristotle's God was separate from human existence and played no role in the universe itself. Apart from serving as the ultimate source of motion this God is completely unaware of anything external to itself.

Aristotle insisted that the material world could not have come into being from another material entity. For if it did, one would have to ask from where did this other material come?

In his fourth book of Physics, Aristotle argued that the existence of a vacuum is impossible inside or outside of our world. Space is always full of matter, which resists the motion of bodies. In the absence of matter in a vacuum, resistance to motion of any kind would be impossible. Without resistance to its motion, a body would move instantaneously, which is impossible.

Kant

Kant believed that while the existence of God could not be proven, we ought to come to a belief in God's existence by way of "logical understanding."

Kant concluded that this world was not sufficient in itself and that an external power, which he identified with God, was a regulative necessity; and that God was a requisite for morality; which gives meaning to our life here on earth.

He also proposed that none of reality exists; reality and all that is in it, including human beings are all part of a dream world. As he saw it, space and time were purely mental intuitions which made our grasp of external reality possible. The substance of thing-in-itself, "Ding an sich", was hidden from human reason -- reality was perceived, rather than led an independent existence. We perceive reality only through the forces, of attraction and repulsion, which work in space.

Gottfried Leibniz

According to Leibniz, real things are substances. A substance is a being which is capable of acting on its own. Each substance has any number of qualities. God is the one and only infinite substance, and through the act of creation, God brought into existence infinitely many other substances.

God created the best of all possible worlds. That is, God considered all possible combinations of substances and chose to create the best one. The choice of the best above all others is in accordance with God's nature.

Perspectivism

Perspectivism is the philosophical view developed by Friedrich Nietzsche and states that we always adopt perspectives by default, whether we are aware of it or not, and the individual concepts of existence are defined by the circumstances surrounding that individual.

This implies that no way of seeing the world can be taken as definitively "true" and that reality is known only in terms of the perspectives of it seen by individuals or groups at particular moments.

Well I could go on and on naming philosophers and religions and telling you what other people have said about life and creation itself but I'm sure you get the picture by now. The thing that amazes me about some of these philosophies and religions is that whenever someone truly stops to think about creation they usually end up at more or less the same place, if you read carefully through some of their ideas or interpretations of the world you will see that there are common threads linking almost every great religion or school of thought. I have barely touched the tip of the ice-berg by naming a handful of them and I would encourage anyone interested in life to explore the subject more intimately.

It is incredibly fascinating that many of the ideas formulated by people using pure thought or reason, long before modern science, bear so much resemblance to what cosmologists and modern day physicists physically studying the world through their powerful telescopes are telling us about the universe and ultimately about creation itself.

I sincerely believe this is because the answers are all around us, and have always been, modern society has become very distracted by technology, we are out of tune with the 'real' world around us and more focused on the world we have created.

Scientists are fooling themselves if they think they have told us anything that has not been said before, sure, the terminology may be different, but have they given us any tangible new answers to the mystery of creation? The ancients talk about spirits who cannot be seen, modern science talks about dark matter and dark energy. Aristotle talks about the "Prime Mover", Stephen Hawking talks about gravity. Dr. Michio Kaku tells us about multi-verses and alternate realities that exist outside our universe, Hinduism and Christianity talk about reincarnation, heaven and hell, places that exist beyond our world.

Our understanding of these forces and concepts may be much better or worse for that matter but that is all it is, there is nothing new. Bear in mind that even when these ancient philosophers and cultures first discovered these things, even they were not telling us anything new, they were simply defining what has always been there. We cannot create anything new inside the universe; we can only understand what is already there.

Ultimately life is a complex and elusive enigma that cannot easily be summed up by any individual philosophy, science or religion. Or can it?

Aside from rational thinking, detailed study and established doctrine I think there are a few other aspects about life we need to consider before we can make any proper or all encompassing theory about the true nature of existence.

CHAPTER 6

Entertainment

Art

Art is not as easily definable as you may think. I could name a few activities we consider to be art such as painting, dancing or music. I could quote the Oxford dictionary once again and define art as "human creative skill or branch of activity concerned with imitative and imaginative design, resulting in visual representation".

I would prefer to settle on saying that Art is an expression of self. No matter what form it takes, if anyone is capable of expressing themselves in any medium. That is art. It is the thing in which the IDEA is expressed.

Sadly we live in an age where art is not pure. These days' artists are rarely motivated by the search for truth; instead we do things more for financial gain or fame rather than a true expression of what we feel. So anyone searching for truth would have to be very discerning about what to consider 'REAL' art.

So why do people do art? Some do it for financial gain, fame or love, while others would tell you that they have no choice, they are driven or forced to do the things they do simply to get it out, to show the world how they feel inside, a select few will tell you that they simply do it for the sake of art, art for art sake.

Whatever the motivation, I think art can only be the product of the mind that produces it. You can't call anything your art if you paid someone else to do it. Art is made in the image of the creator and is a visible or tangible transformation and expression of the self.

It is a misconception that art is simply trying to imitate life, if that were the case anybody could be a great photographer, you simply pick up a camera, take a photo and you have captured reality. But that will not necessarily produce a great or artistic photo, to be truly great the photo would have to contain or reflect something more, something deeper, that something special can only come from the mind of the photographer. Real art must have an extra quality or deeper meaning.

Why is it that an exact duplicate of the Mona Lisa is worth so much less than the original? The original must contain some essence of Leonardo himself; also, the original is original.

So art must be an expression of the self and art must be original or unique in some way. What else?

There is a famous legend concerning the artist Giotto di Bondone who was asked to demonstrate his skill as an artist to the Pope. To do this Giotto drew a circle so perfect that it seemed as though it was drawn using a compass.

I think this story is hugely significant when it comes to explaining the true nature of the universe. Not only because Giotto uses a circle to represent perfection or great skill but also because such a simple story has survived for hundreds of years. The answers to the universe can be found all around us. It is in the things we do and it is in the things we treasure, we are drawn to certain things on a subconscious level because these things define and represent truth.

Fiction

What is fiction? Why do people write fictitious novels, tell farfetched stories or create unrealistic movies? Why do we spin yarns about romantic characters and send them on dramatic, often exciting adventures? Is it a need to escape reality or do we use it as a way in which to transform or spice up the mundane?

Like any other form of art I think fiction is basically a form of self expression, we can use a story to tell the world what we think or to pass comment on society, but I think any work of fiction has an added quality to it, it has something extra, and I think that something extra is experience. When we create a work of fiction we can experience things that we do not necessarily experience in 'REAL' life. I am sure many writers live vicariously through the heroes and villains they create.

You may be wondering what any of this has to do with the universe, well, like I said earlier, I think the answers to the universe are all around us and can be found in everything we do, we simply need to look closer.

To quote Shakespeare:

"All the world's a stage, and all the men and women merely players: they have their exits and their entrances; and one man in his time plays many parts, his acts being seven ages".

Cinema

Probably the most entertaining form of art in modern culture is the cinema. When it comes to experiencing exotic locations and adventures the celluloid world of Hollywood has few rivals. So why is this form of escapism so popular? What makes a movie great?

The first thing any movie-buff will tell you is that it is the acting. A great movie usually has believable actors. And why is this so crucial to a film's success? Because a great actor can make you feel all the emotions they are experiencing, they make you laugh when they laugh and they make you cry when they cry. You can live out the story through the characters you see on screen.

Aside from great acting and a riveting story line a good movie needs a realistic setting, if you look at some of the most popular movies, you will find yourself immersed into its spectacular environment and setting. With advances in special effects and computer generated imagery you almost believe that you are seeing a brave warrior running through an alien jungle on a planet with 3 moons or that you are aboard an airplane that is about to crash into the ocean. The attention to detail and the professional quality of the end product is so realistic it is almost as if it were REAL.

Yes, when it comes to entertainment and story-telling few things can compete with a thrilling and exciting movie, until recently that is. Over the last few years we have witnessed incredible leaps in technology. And in one particular field these massive strides in technology is becoming more and more popular and entertaining and even threatens to surpass its cinematic rivals. That is the virtual world of computer games.

Computer Games

If you haven't played a console or computer game within the last few years than you are in for a treat. With the tremendous advances in technology people are now able to immerse themselves more and more into these incredibly realistic environments. Why go and watch a movie where you have to rely on the actor's ability in order to truly appreciate the story when you can virtually live through the entire ordeal on your own.

It is truly remarkable to witness the amount of detail and spectacular realism that is invested into creating these games and virtual worlds and I am sure that based on the rate of advancement in technology, it will only get better, more incredible and more REAL as time goes by. I am convinced that one day you will be able to place a gaming helmet on your head, or even swallow a 'blue pill' and not know the difference between reality and fiction. You will be able to feel every breeze, listen to every little cricket chirping and probably experience every tiny little mosquito bite. Imagine where technology will be in only a few hundred years, now imagine where we will be after a billion years. This is where people are lacking in their search for God; they cannot follow a single line of thought through to its logical conclusion.

What are the ingredients for the ultimate in entertainment? A good story, a stunning environment, a way in which to truly experience every aspect as though it were real and most of all, a challenge. Why do something if it was not going to be challenging or riveting, if there is no thrill there is no point.

Am I saying that the universe is a computer game in which we only have one life in order to up the adrenalin rush? No, but if you find that scenario incredulous wait until I tell you what I really think is happening. If the universe were simply a giant computer game then once we die we would still wake up into TRUE reality and be faced with all the same questions we have today. Anyone who has seen the Matrix movie will understand that once Neo is awakened and is unplugged from the machine he has been connected to all his life, he still has to face a reality similar to the physical real world we experience right now only machines now rule his world, but he is still a mortal man inside the standard universe. So what is the universe really?

CHAPTER 7

Dreams and the Occult

Dreams

Everyone dreams. There are hundreds of books on dreams and what they mean. I honestly don't think I have to tell anyone what a dream is. People have always dreamt and tried to interpret what their dreams mean.

I think the sole purpose of dreams is to provide you with the clues you need to answer the question of what the universe is. The dream itself or rather the fact that we possess the ability to dream is more important than the content of our dreams.

When we dream we are submerged into a totally new world, a world we have created, we experience many of the sensations that we face in real life and most of the time we are hardly aware that we are in fact dreaming, that is how real and lucid this fictitious dream world is. In dreams we can even conjure up people that are long dead and they will appear so real and lifelike that we are totally convinced we are in the company of those actual people.

I don't want to be dismissive of the content of dreams but I really don't think they are portents of the future. I think dreams are there so that everyone can understand the true nature of reality. I don't think the world is there only to be understood by scientists who have the advantage of studying the stars or black-holes, I think the tools we need to answer the great questions about the universe are all readily available and require only a rudimentary scientific and mathematical knowledge. Anyone can have a dream and wake up, only to ask themselves "how could that seem so real, what is real? Is this world real or is it only a dream?'

A dream is the realization of an idea or thought, albeit so elusive. In a dream we can create reality out of thin air and we can even create living creatures. When I was a kid I longed so hard for a bicycle that I dreamt I had one, it was so real and tangible that when I was awaking into the real world I tried desperately to pull this bicycle out into the real world. No point telling you how that ended.

Let me reserve the right to say more about dreams and their supernatural aspect or ominous nature later.

The Occult

The Occult is a study of 'hidden wisdom'. Many people believe that there is a deeper truth hidden beneath the surface of what we see. Astrology, magic, numerology and even alchemy are examples of the occult.

Whereas science is normally concerned with the exterior nature of a thing and trying to explain or define a thing by its physical properties, the occult can be said to be more interested in the deeper meaning inside everything we see.

For example, physicists would study the cosmos and tell us how the universe works due to the physical laws of nature. Science will tell us that the planets orbit the sun because they are bound by its gravity. Astrologists on the other hand will be more concerned with the hidden reasons or deeper meaning behind the placement of the planets and stars at any given time.

I would define the occult as a study of the notion that everything in the universe happens for a reason. I only mention the occult because of its popularity throughout the ages and the huge effect it has had on the development of modern society. It is thought that Hitler himself was a big believer in the occult and that many of his policies were based on prophecies interpreted to him. If you look at the consequences of these beliefs you can see the great effect they have on society. I have even heard of wealthy, well educated modern day investment bankers and brokers investing in the stock market based on astrological predictions and omens.

Some fortune tellers or witch-doctors can roll a bunch of bones and claim to predict your future based on the outcome of the final position of the bones. Others find meaning in tarot cards or palm reading. During the 2010 Fifa Football World Cup many people were placing bets on the outcome of the matches based on the predictions of an octopus named Paul.

Because of the negative image associated with the occult many people would not openly admit to believing or subscribing to this element of society. And yet those same people will casually tell you that it is bad luck if a black cat crosses your path or that you will have seven years of bad luck if you break a mirror. If something good happens to them they will say things like "it must be my lucky day". I think on some level we must all harbor some kind of innate feeling that everything around us is happening for a reason. I don't think anyone truly believes that life is simply random.

CHAPTER 8

Instinct

What does instinct tell us about the universe? Instinct is the intrinsic qualities hardwired into our very being that tell us how to act in a given situation.

I recently watched a documentary regarding human instinct and it was fascinating to hear stories about people who when faced with extraordinary circumstances act in remarkable ways without even thinking about it. There are times when we will simply do something without even considering the consequences.

I want to know what our instinct tells us about creation. What do we instinctively feel about life? This is a difficult question because as soon as you think about it you are overriding your instinct and using your intellect. Without sounding condescending, I think we can answer the question by looking at the demographics for the earth. Most people on earth believe in a God of some kind. And they believe that we are here in this life for a reason. Why do they believe this? Is it because that is what they were told when they were children? Is God's popularity a question of nurture rather than nature? Or could God be the product of pure instinct?

I can propose a hypothetical situation in which a couple of people with average intelligence wake up on an island with no recollection of any previous life. What will their instinct be? They will obviously eat and drink in order to survive. Those will be their natural survival instincts. But what would they think about life? They would see other creatures around them, some would kill each other. They would experience death; they would see how something could be alive one day and dead the next. They would witness deaths corrupting influence throughout nature.

What possible thoughts would go through their heads? What would they think when they lay and look up at the stars at night? Will they feel like they are being watched? What will they dream? If one of them gets injured or sick will they worry that they might die like the creatures and plants around them? If they mate and have a child, what will they think about life then? Will they think it was all magic? If one parent dies when the child is older, what will the surviving parent tell the child about death? Will they naturally start to wonder why? Why are they here, what is this all about, where did the dead person go? Is the decomposing husk all that is left of the person they knew? Why does it not move and laugh and cry like it used to? Surely the question of WHY must be instinctual. That is the one common thread in all of philosophy, science and religion; everyone is wondering or trying to answer why. But what of the answer, what is the natural instinctive answer?

What would you tell your child in such a situation? If you are of average intelligence you will start to study your environment for answers or clues as to what is happening. You will see that over time everything dies and that new things are born and grow in its place. You will see a pattern to life. When you look up at the stars you will see similar patterns. Surely the answer must be here somewhere. As elusive as the answer may be, the question is natural and instinctive. What is life? What is happening?

And what of the people who say it is all just random. Could it all possibly be random? Is it just good luck that the earth is not any closer to the sun or we would roast, is it coincidence that we are not further away from the sun or we would freeze? Is it great fortune that an errant comet or meteor struck the earth millions of years ago and killed the dinosaurs or else we would never have evolved? All random things happening for no real purpose. Did your parents meet by accident, did theirs? If you think of every little coincidence that led to your being here right now, is it all just lucky timing? What else could it be? I guess it could just be an unlucky coincidence that we cannot see behind the curtain of cosmic microwave background radiation to the exact point where creation began.

A good way to explain the universe or yourself as a random result would be the analogy of a deck of playing cards thrown up into the air. If you took a deck of cards and simply flung them up into the air they would land in a random scattered pattern, a 4 might lay near a 5 while a 7 lands on top of a king, all random. Now, if we take a snapshot of this random scattering of cards and call it the ideal outcome or profess that this was meant to be, no matter how hard we tried, if we kept throwing the cards up into the air, chances are they will never ever land in the same exact pattern within our lifetime, in other words, we cannot look at the end result of a random event and call it ideal or destiny, simply because we can now never repeat the exact same scattering of cards we achieved in the first place it does not mean that first arrangement was ideal, we only said it was. We cannot look at the world as it is today and say that everything was conspiring to make this world exactly as it is; this is simply a random result and not the result of any conscious effort. So are these people right? This is certainly a strong argument in their favor.

I don't know what my instinct is about life but I can tell you right now what my logical conclusion is about the universe being a random event. The answer is of course NO. The universe is not random, and I'll go one step further to qualify my position.

When it comes to random events like the example of the deck of cards flung into the air and not landing in the exact same position the important thing to remember is that on an infinite time scale not only will the cards eventually land in the exact same position, it will land in the same position an infinite number of times. On an infinite scale the cards will always land in that same position.

It amazes me that people are so willing to latch onto the idea that there are an infinite number of universes out there, but they call our universe random. Or lucky. But logically it follows that if there are infinite universes out there. There must be an infinite number of copies of our exact universe. So is creation lucky? Or divine?

CHAPTER 9

Me

I feel it is imperative to state that I am nobody. I am literally some poor unknown person from an insignificant suburb in Johannesburg, South Africa with limited formal education.

Yet, I am going to tell you something different about the universe. And the very fact that everything in my life led me to see the world the way in which I do leads me to believe that life is not random. I have a perspective on life that only my particular circumstances could have created. Whether or not I am right about the nature of creation you will have to decide for yourself. I can tell you it is incredible and it is *unique*.

I have written two previous books. Though I must warn you, I am impulsive and very emotional and passionate when it comes to life. Due to my circumstances I have very little faith in established organizations and society in general. The first book I wrote is entitled 'Terms and Conditions' and basically tells the story of my life and the search for my place in society. It's fair to say that the book did not set the world on fire and will probably fade into obscurity.

On a personal level, writing that book was a crucial part of my journey to what I feel is 'enlightenment' as it allowed me to understand exactly who I was and what was important in life. I have little concern for my ability as a writer. I'm sure anyone reading it will appreciate what I was trying to say, regardless of my grammar, spelling and prose. I am more interested in sharing my ideas.

'Terms and Conditions' reflects a time in my life when I was indifferent or blasé about the existence of God and more consumed by the day to day struggle of life.

The second book I wrote came about after I started to see the world from a different perspective. Once I understood how society worked it became easier for me to see what motivates people and what drives them to do the things they do. I was no longer consumed by the day-to-day struggle for success or recognition. I had already abandoned all faith in society. Now I wanted to know more about the universe and what life was really about. Amazingly I discovered that even here, in this grand arena, people were too afraid to step outside the box of conformity. I wrote a second book entitled 'Eternal Life, The Universe and Happiness'. In this book I theorized that people cannot die and that they in fact exist in their specific space and time for all eternity. Naturally this type of thinking can only lead to a direct search for God.

Once again, the physical process of actually writing this second book allowed me to see the Universe from a very different light. I feel I have uncovered so many answers that have really been there in plain sight all this time. In my excitement I have slowly been going crazy with the information now residing inside my head. I have been trying desperately to tell the world what life is all about. I am using all the limited resources available to me, i.e. the internet, publishing services, or even simple word of mouth.

As anyone who has ever tried will confess, getting people to listen to you is not as easy as it may seem, especially if you are a nobody.

So, if you are reading this, I want you to know that getting this message to you was not an easy task and you must understand that this struggle is truly a part of the message itself, if I am right; this message is proof that everything and everyone in life does indeed have purpose.

Okay, allow me to tell you what I think the universe is and what exactly is happening here. It may not be absolutely right but I feel it is the first 'ALL ENCOMPASSING' explanation to the riddle of creation and at the very least it is not conventional.

You, the reader, have every right to remain skeptical about what I have to say, my intention is only to share my ideas and to get you thinking about creation from a different perspective.

PART II

THE ANSWERS

CHAPTER 10

Special Relativity

Much has been said about the sheer genius of Albert Einstein. What most people don't realize is that it was not Einstein's great mathematical ability or scientific knowledge that qualified him as a genius. It was his uncanny ability to see the world in a totally different and unconventional way that has opened up the doors that now allow us to explain so much of how the universe works in a totally scientific way.

Yet, as much as I admire this great man, I believe the answers to the universe are more than merely scientific. And, even though it was Einstein who gave us the theory of Special Relativity, we must remember that these laws of physics were always there, Einstein merely helped us to understand them better. We don't invent the laws of the universe, we only discover them.

As mentioned, one of Einstein's greatest contributions to science was his theory of 'Special Relativity'. But what exactly is special relativity:

> "The physical theory of space and time developed by Albert Einstein, based on the postulates that all the laws of physics are equally valid in all frames of reference moving at a uniform velocity and that the speed of light from a uniformly moving source is always the same, regardless of how fast or slow the source or its observer is moving."

Okay, but what does this really mean and why is it so important to understanding how the Universe works or whether or not there is a God?

I'll give you my simplistic understanding of special relativity and tell you what it means to me. Special relativity says that any object travelling in space, regardless of its speed, will always see light travelling at a constant speed and that time will always be relative to each observer or that time will effectively move differently for each observer depending on their speed.

This has two major consequences. To better explain this, let us use an example of two spaceships travelling in space. In the first scenario assume that the spaceships are both standing still or rather hovering and facing each other in space, so they are face to face with maybe a few hundred feet between them. If the one spaceship fires a missile at the other ship, that missile will travel at a certain speed. Fine. Now imagine the same situation, only in the next example, the one ship is standing still but the attacking ship comes flying towards the first ship at high speed and then fires the missile. In the second example the missile will move much faster because it will have its own speed plus the speed of the ship that fires it. Makes sense.

Now, reset the scene. Two spaceships face each other and both are standing still, if the one ship turns on its lights, that light will travel to the other ship at the speed of light. Once again that makes sense. But the curious thing is that in the second scenario, if the second ship comes flying through space towards the stationary ship at a super high speed and then turns on its lights, that light will still only travel toward the stationary ship at the speed of light. It will not inherit the speed of the ship that emits the light as did the missile. This means that you cannot add to the speed of light. Light will always travel at the speed of light no matter how fast the object that emits the light may be travelling.

That is one aspect about Special Relativity. The second is that time will always be relative to the observer. Use the same setting and this time place a big clock in the window of each spaceship. If one ship is standing still and the second ship comes zooming past at phenomenal speed, and assume the person inside the stationary spaceship could see the clock through the window on this speeding ship, for the person standing still the clock inside the moving ship will appear to move much slower than his or her own clock. Time will appear to be ticking along much slower aboard the moving ship. But, for the person aboard the speeding vessel his/her clock will simply appear to be ticking along as normal. And strangely enough, for the person on the speeding ship looking out at the stationary ship, it is the clock aboard the stationary ship that seems to be ticking slower. Time will be relative to each observer.

An astonishing feature regarding this phenomenon is that of time dilation. Remarkably, if the person on the speeding ship can travel at very high speeds, that person will age much slower than the person standing still, so although time is relative to each observer, time seems to have less effect on the person moving at ultra high speeds. This is a phenomenon that baffles scientists because as I said, for both people in this example it is the other

61

persons clock that appears to be moving slower, yet it is only the person that is speeding up who actually ages slower in comparison to the one standing still. This is often called the "Twin Paradox" and remains unexplained. But I believe I have the answer to this riddle.

Okay, that was a very simplistic explanation of what special relativity means. But you know what, Einstein did not invent Special Relativity, he simply defined it or discovered it. It was always there. We only need to look at the world more closely to uncover these things. We have to show initiative and focus on the things that really matter. If you think that special relativity and time dilation does not concern you, then you are very mistaken. Time dilation is what makes your GPS system work.

I only started taking a keen interest in what Einstein had to say after I started thinking about creation and my place in the universe. I was trying to figure out if there really is such an entity as God and what my purpose was here in this life.

Naturally I began to wonder about death and time and what happens after a person dies, should I spend my life and energy working hard to be happy or should I steal money and get rich quick so I could buy all the happiness I could before I die, will God judge me?

Amazingly, as I began to focus on the true nature of the universe I began to see things in a very different light; the first thing I discovered is that time is a very peculiar concept. How can civilization have come this far without really having understood the concept of time? We have been around for so many generations and nobody really understands what time is. What have people been doing with their time?

You know what is even more astonishing? Forget about the generations who have lived long before us, amazingly even after Einstein came along and gave the world his theory of special relativity, people still do not realize what this theory entails. The scientists we put so much faith into simply came along trying to prove or disprove Einstein with elaborate experiments and tests, but nobody really understands the massive implications of what Einstein was actually saying.

The deeper ramifications of Einstein's theory of 'Special Relativity' still eludes us. And no, it does not surprise me that we can develop GPS systems and find ways to make a commercial profit from Einstein's work yet still not truly understand it. But I think Einstein knew or suspected more than he was saying. I like this quote of his that I used on the cover of this book. "Reality is merely an illusion, albeit a very persistent one".

Time is relative to the observer. I discovered this for myself long before I understood Einstein's theory of special relativity. I simply started to wonder how it was possible that I could be alive one day thinking that the entire world revolved around this one moment and that the present time meant everything and then the very next day I would find my own previous behavior and actions so irrelevant. How could something be so important one day and not matter the next? It's because the next day would now become so important of its own accord that I simply had to focus on this new moment in time. Time was only relative to me on each new day.

The more I pondered this question. The stranger my finding became. I started to wonder what would happen if I could travel back or forward in time. Would I still think certain things so important? Thinking about time travel started to raise some very tricky questions. My only understanding of time travel was what I had seen in fiction like star-trek or superman, if you could fly faster than the speed of light around the planet earth you could go back or forwards in time. For me this made no sense. If I am here now how could I go back to the past and be there as well, this would mean that there had to be multiple versions of me. And there it was. An answer to time that nobody has really understood before. Since then I have read a great deal about Einstein's theory of Special Relativity and I have read a few books by Stephen Hawking and other popular physicists regarding the subject. I now don't believe that people will ever be able to travel through time. Neither forward nor backward.

Yet, the simple realization that to even theorize about time travel would mean that I would have to exist in multiple places at different times opened up some of the greatest revelations I have ever known. And amazingly enough, in all the subsequent books I have read, I have not come across anyone else saying the things that I am about to tell you.

As I mentioned earlier, I may or may not be right, this is almost inconsequential. The fact is that I have seen, or believe I have seen something different.

Time is an illusion. For me to be here right now, with this old scar on my arm where I got injured falling from a home-made go cart as a little kid, I would have to be in several places at different points in time. I must be there in the past, doing all the things I have done, in order to be the person I am here in this present moment. Cause and effect. I must therefore exist in multiple places at different times. Because if I could theoretically go back in time to that moment I would see myself there as a little kid, all of the past must be preserved or stored somewhere. The past must be as real as the present. Do you know what this means? It means that if the past is real, and the present is real, both must be equal and equally relevant. All of time must be relative to the observer.

If I could go back in time I would find a little kid building a go cart and thinking that this was the only moment in his life that mattered. As I sit here now, writing this, I feel like now is the only relative moment because everything is so real and physical around me. But every person at every point in time will feel exactly the same about every specific moment in time, you will call it THE PRESENT. Everything will seem real around you. All of space and time must be relative to the observer exactly as Einstein's theory implies. What people don't realize is that this applies to yourself at different points in time as well.

The people who came after Einstein didn't really bother to truly understand what this means, they dissect Einstein's theory and study its physical application but they don't understand the deeper meaning. Time is relative to each observer even if you yourself are the different observer viewing time from different points in space. No matter where you are in time and space, NOW will always feel like NOW and your clock will tick normally. Because I feel I have a deeper understanding of what this means, I can not only rule out any form of time travel to the past but also to the future.

Even if you could somehow accelerate yourself to almost the speed of light and age much slower than the people you leave behind (time dilation) when you stop your vehicle and find that it is five or ten years later and all your friends have aged faster than you, you will still only be in the present. That new time will be your present; you cannot physically stand in the present moment and say you are in the future. If this were the case one could cryogenically freeze oneself for a hundred years and then get revived and say that your cryogenic chamber was a time machine.

You must look deeper. Even now I bet the true meaning of relativity escapes people. If you are alive at different points in time and space and each moment is relative to you at that specific moment. You cannot die. You will remain in time and space for all eternity. If you get killed tomorrow, that does not erase your past. If I could go back to your past even after you have been killed I will still find you there doing whatever you were doing at that time, the past is what creates the universe we see today.

When we talk about the universe as a whole we cannot look up into space with our advanced telescopes and take a photo of the cosmos and say that this is the universe, what we see is only one moment in time, the universe itself is something that has been happening over billions of years. For this very reason we cannot think of the universe as a fixed place, the universe is an event, it is something that is happening or unfolding even as you read this.

Einstein's theory of special relativity tells us that no two people viewing an event from different points in time and space can agree on when exactly that event took place.

Scientists tell you that special relativity means you will see clocks moving at different speeds and they give us GPS and satellite navigation systems as a result but there is a deeper message, if special relativity is right, it means that you will always be alive in your space and time, thinking that THIS is the present.

Now, if you really think about this, forget about the commercial aspects, really think about it, this message was always there, Einstein did not invent special relativity, so anyone living thousands of years ago could have understood this if they really looked closely enough, and where have you heard something like this before, a promise of eternal life? Many religions talk of an eternal life and they tell us this long before science.

3 books later I have learnt not to disregard everything people believe simply because you think it comes from ignorance and delusion. The answers to the universe are all round us. Maybe people have simply misinterpreted the information.

I don't subscribe to any religion; I think that like science their views are too narrow minded or misguided. Religions tend to segregate and divide people whereas if there truly is a God, everything in this universe is connected and is a part of God. I might sound like I am negative towards modern science, but in fact I think the pursuit of scientific knowledge is the highest calling. Because one day, science will in fact prove the existence of God. But science has to remain open-minded and not conform to popular believe as though it were religious doctrine. Science must be willing to listen to new ideas.

Amazingly, in my search for answers, I found that both Einstein and Isaac Newton, two of the greatest scientific minds that have ever existed, both felt that time was something of an illusion and that somehow, the past, present and the future all existed simultaneously.

So what else is happening here?

CHAPTER 11

Black Holes

What is a black hole? People think that our modern civilization is so much better than those that have come before us because we can build giant telescopes and send them into space to study these giant monsters that eat stars. We think we are great because we can see great things. We forget that discovering something does not equate to inventing something. Black holes or whatever they are have always been there.

Despite having read a great deal on the existence of black holes and having watched one documentary after another by famous tele-physicists on the discovery channel adamantly proving the existence of these invisible monsters out there, vehemently professing that all the scientific data points to these denizens of the dark swallowing up stars. I will now tell you that from my point of view there are no such things as holes in the universe.

Like 'Special Relativity' a black hole is a term we have coined to describe some phenomena we discovered in the universe. And even though I am not a formally educated scholar I will show you the value of truly understanding something. Once this book is done I will struggle to get it published because I do not come from some highly decorated college or university, publishers will look at my prose, or my grammar and spelling and reject me as unworthy of publishing or not financially viable. As with everything else, the true understanding of language and what it is meant for has become lost. In order to be heard you have to be profitable and marketable. It doesn't matter if you write volume after volume of second rate fiction about vampires and werewolves that attend high school, as long as you write it according to the commercial mainstream rules, they can sell it. But you will witness the true power of language once you see how the simple definition of something like a 'black hole" can change the world.

There is no such thing as a literal black hole in space that stars and planets are falling into. This concept has become popularly accepted because of sheer misunderstanding and misinterpretation. Matter is not falling down a black hole and leaving this universe. Nothing ever leaves the universe because the universe is not a place. The universe is something that is happening. And 'black holes' are a part of what is happening.

Let's go back to Einstein; Einstein said that nothing could travel faster than the speed of light. And this is where the misunderstanding starts. Einstein said it and the result was "let's build a Large Hadron Collider and try to accelerate a particle to go faster than the speed of light to test it". But what does it mean to say that nothing can travel faster than light? Why don't we try to understand it before we try to prove or disprove it? Why can't anything move faster than light?

When we see something in the physical world we see it because of the light waves bouncing off or being emitted by those objects. If I look at an apple I am seeing the apple because of the light it reflects. If we look out into deep space and we see stars shining brightly, we are seeing those stars because of the light they emit. But what is happening? When we see something we are seeing it because of the light but we are able to define or interpret what we are seeing because of the information carried by the light. Nothing can travel faster than light. I would elaborate on that by saying, information cannot travel faster than light.

The Universe (The Event) is happening at the speed of light and nothing can happen faster than it is in fact happening.

Another thing to note is that Einstein does not say that nothing can travel at the speed of light; light travels at the speed of light, he says that nothing can travel faster than the speed of light. So in fact information can travel at the speed of light but not any faster.

And what is information? Information is the stuff that defines or characterizes a thing. Information is the coding inside the matter. Information is intelligence. When I see an apple, the information that is being carried to me on the light is that it is a small round object of a red or green color. The information is what defines the object. My brain processes this information and I see the apple. It is the same with everything else we see, the information is carried to us by the light and we define what we are seeing based on the information or coding. The information is the essence of the thing.

Information can travel in other forms as well, for example via sound waves. If we are sitting in the dark and we hear a window break we can normally associate the sound we are hearing with the breaking of the glass. In this case the information travels much slower because we are not seeing the window; we rely on the sound that travels to us via the speed of sound. Think of another example in which we see a super-sonic jet go tearing through the sky. We see the jet long before we hear the sound because the light information that tells us this is a jet reaches us long before the sonic boom which travels much slower. So information can travel in different ways but the fastest way that information can travel is at the speed of light.

Now, with this deeper understanding of the speed of light, what does it tell us about the universe? The answer comes from truly understanding what 'Information' is. The information we receive via the speed of light is the very definition/essence of what the thing is. The information is really the soul of the thing. Information describes and defines an object. Programmers can think of it as the code behind everything we see on a computer screen. So, what are we really saying when we say nothing can travel faster than the speed of light? We are saying that no object can travel faster than the stuff that defines it. An object cannot travel faster than itself. An object cannot travel faster than its soul. Therefore under normal conditions we will never be able to travel faster than light. Do we really understand this? I don't think we do, or else we wouldn't be wasting millions of dollars trying to prove it. Though, I suppose proving it has its benefits.

Take the example of the supersonic jet, if this jet could travel faster than light, what would we see? Assume that the jet is now

24 hours faster than the speed of light, so the jet is in fact 24 hours faster than the very stuff that defines what it is, when the jet reaches us we would not be able to see anything, maybe we would feel the gravitational pull of the jet, but we would see nothing, the jet itself would simply be pure imperceptible energy, a day later the light (information) would arrive, so maybe we would then see a ghost of the jet but no real physical jet will be there, it will be 24 hours ahead. Does anyone still think we can ever travel faster than light? Please explain to me what you think will happen. Will we simply know a jet is here and then see it the next day? What would be the sense of that? Where is the logic?

So what does all this have to do with black holes? And why do I say there is no such thing as a black hole? And once again, what has any of this have to do with God?

Well, a black hole or rather this super dense thing with its incredible gravity that we are experiencing at the centre of almost every galaxy. This is not a hole. It may be black or invisible but for a totally different reason. I say that the thing at the centre of the galaxy we call a black hole is actually an object that is moving at the speed of light. It is not a hole at all. It is simply a non-descript mass of super-dense nothingness. Or pure energy if you will. It is something that is travelling so fast that it has separated from the 'Information' that defines what it is. A black hole is pure unidentifiable mass or energy that has split from its coding or soul. A black hole is the place where Einstein's $E=MC^2$ is realized. This is where the energy / mass equivalency is reached. This is where creation unfolds and is broken down into its basic building blocks.

Einstein deduced that energy and mass (matter) must be different forms of the same thing and that the two were totally interchangeable. Energy can be converted into mass and vice versa, all you need is movement. But we also know from the first law of thermodynamics that energy cannot be created or destroyed, energy can only be converted. Nothing is ever lost or destroyed inside the universe and nothing ever leaves this universe. There is no such thing as a literal hole in space through which matter is falling.

When physicists study a star or other object that gets closer to a 'Black hole' they see that the object is accelerated to phenomenal speeds and then it suddenly disappears and is swallowed up by the black hole. Well, this is not really what is happening. As the star or planet gets closer and closer to the 'Black Hole' it is being accelerated faster and faster until it is moving so fast that it reaches the speed of light and is separated from the information that defines what it is, this nondescript pure energy then merges or joins the super dense nothingness while the information or soul of the object is scattered or preserved inside the universe.

Remember the universe is not a place; the universe is something that is happening. So what is happening? The universe is like a great engine or production line for $E=MC^2$. Energy is first converted into mass and then this mass is later converted back into energy. All happening at the speed of light.

We cannot see a black hole because it is pure energy. It is not even black. It is simply a super mass of gravity or energy with no identifiable characteristics. While the information or soul of the object has been split from it and is left on its own. There is only the idea that there was once a star or planet or object. There is no loss of energy or matter because the matter has not left the universe down some black tunnel. The object is likely converted back into pure energy.

That sounds incredible right? I bet any publisher reading this would be terrified of publishing such an outrages idea. Their organization would be the laughing stock of society. Not because this theory is so incredible though, only because I am nobody. What could I possibly know about black holes? Instead they will publish book after book or produce one documentary after the next telling people that black holes are portals to alternate universes in which matter is swallowed up and transported away, in this alternate universe there are other versions of you living life as a pink elephant or you might be able to live forever in this alternate universe because the laws of physics are so different.

Of course they would be right to be skeptical because I really don't know anything for sure about black holes, I only know what people long before me knew, people like Plato or even Jesus, and that is that on some level we exist in two different forms, there is the idea of us and there is the physical manifestation of us.

Einstein knew it, Einstein never had access to the powerful telescopes we have right now, and he didn't know for sure that black-holes would exist. But he knew that at some point in the universe there had to be something like a black-hole.

Society needs to change. It is shocking that Einstein was widely ridiculed when he first proposed 'Special Relativity' because it went against everything Newton had said about the laws of physics. But a man must be brave enough to take a stand against conformity if the world is to change. People must realize that being different does not make you a fool, it is the very institutions that simply do something one way because that is the only way they know how, who are the fools. It is the capitalists who are holding us back.

I'm sorry but I have to use this message to change society. Because I know that I am right and that one day this will all be proved. I want people to know that they should all live their lives to the fullest regardless of how society views them. There is a place and a purpose for all of us in this universe. If there is a purpose for every little person, there must be a higher purpose.

So what does all of this have to do with God? Well, one of the biggest problems I have had with the religious interpretations of the Universe has always been that I could never reconcile the fact that a person cannot exist without their physical body. How can I still be around after my body dies? Well, if my theory of black holes is correct and I were to be captured by one, my physical self will be accelerated to the speed of light and I will join the great *Energy* while all my information, my SOUL, will be left behind here in the universe. The idea of 'ME' cannot cease to exist. The universe is eternal.

And guess what, if there is a 'black hole' at the centre of every galaxy, nothing is going to stop it, it will keep growing and growing until it consumes the entire galaxy. Eventually the entire universe will be split into pure energy and information. Matter and soul. Are you starting to see the beauty of it all? It is incredible. This is how we will always exist.

I think the soul of an object can be defined as what it was/is. I don't think the human soul is a separate entity that can live on or experience new things after you die, but rather the human soul can be defined by how a person spent their life and their time here on earth. The soul is the sum of your deeds. It's what you were/are.

That is why an original painting by Leonardo Da Vinci can be considered so valuable. It has a part of his time and his soul invested in it. His energy was used to make it. The original Mona Lisa is a part of Leonardo Da Vinci for all eternity.

Go back to Einstein's famous equation:

$$E=MC^2$$

$$ENERGY=MASS \times LIGHT^2$$

$$ENERGY = THE\ UNIVERSE$$

The universe is created at the speed of light, this is the universal speed limit, nothing will ever move faster than the speed of light because that is the speed at which matter is defined. Once we accelerate objects to the speed of light, creation itself becomes undone. At this speed an object is split into its basic building blocks of pure energy and the very essence or coding that defines what an object is.

A black hole is the speed at which the universe happens.

CHAPTER 12

The Universe

What *exactly* is the Universe? To answer this I think we have to go back to the question of Art. What is Art? Or more specifically what would be the best or perfect form of Art?

I would say REAL Art is the TRUE expression of self. If a painter paints a picture, that picture is created from his mind or his will. It is an expression of some part of his soul/his essence. But what else is happening? If we look at the physical process we find that the painter uses a canvas to paint on and uses different types of paints, this is the physical material that he will use to express his thoughts; he will also expend some energy actually painting the picture. This energy he will derive from eating food and drinking water. I think we can see a little microcosm of the universe at work in this process.

What would be considered the perfect work of art by a painter or any kind of artist? If you know anything about art or if you have tried to paint something yourself you will know that the artist rarely considers his or her work to be complete. It is never perfect; they always want to keep adding a brush stroke here or a dab there. So what would be the perfect expression of self? I think if the artist could pour his or her entire being or self out onto the canvas and still somehow maintain the ability to keep changing his/her work that would have to be considered perfect, if the artist is able to convert his / her entire self into the art, he she will be finished. The art will now be a true transformation and perfect reflection of the artist, once we see such a work of art we will automatically know who the artist was from every stroke of the brush, the two will in fact be indistinguishable. Not very practical or achievable by human standards in the real world. But what about God?

This is what I think has happened/is happening. I think before the universe there was only pure energy. (**NOTE**: There is no "before the universe" as I will explain later). This is the substance that the universe is made from. And separate from this was the supreme thought or Idea of the Universe as we see it. Or we could simply think of the energy itself as being intelligent or having purpose.

Before the universe is created there is only energy. We know that energy cannot be created or destroyed (First law of thermodynamics) Energy is eternal. Energy does not experience time. Does this description of energy sound vaguely familiar? Maybe ancient cultures simply called energy God. Now, If God (The Energy) does not create the universe there will be no such thing as time.

So the raw or pure energy is there and the Supreme Idea (God) is there. Probably one and the same, the energy has purpose, it is intelligent. From this point forward think of God and energy as the same thing. God then decides that he wants to create the universe. Why? Why would God want to create the universe in the first place? Because thought is only thought, an idea in itself is only an idea, it is not a real thing, the thought of being alive cannot compare to actually being alive and being able to take a deep breath, the thought of a juicy peach is not the equivalent of physically sinking your teeth into one and savoring the juices that burst into your mouth. The thought of love is lonely and sad. The thought of sex is unsatisfying. The thought of swimming or flying or hunting or laughing is not real. So God decides that he will create all of these things and so much more.

Suddenly the unidentifiable energy starts to move; energy itself has no form and takes up no space. Now, when this energy starts to move at the speed of light, it starts to convert into mass, mass has form and takes up space, mass also experiences time. This is why we see the universe suddenly bursting into existence 13.7 billion years ago. It is not as some theories predict that the universe is created by mass leaking into our universe from an adjacent universe and it does not occur from a big bang as popular theories claim. The entire universe does not burst into existence at one particular point in space, but rather at one point in time.

The universe appears as energy transforms into mass. But there is one crucial thing about energy that I have not yet mentioned. We know energy is eternal but energy is also never still. Energy is always moving, energy flows. There is no sudden point in time when energy starts to move. Energy must always be moving. What does this mean? Well, if energy is always moving, and energy is being converted into mass and then back again, it means that the universe is constantly recycling. If energy is eternal, the universe must be eternal. There is no real start of time or end of time. There are only infinite cycles. Our current universe or cycle started 13.7 billion years ago. Only we and physical objects like planets or stars (mass) experience time. Energy does not experience time. There is no "before the universe begun". There is only a start of each new cycle.

God creates the universe in order to realize the idea. This universe is created in the image of God. It is made from the essence of pure thought/energy. Every little thing inside the universe is a reflection of the supreme idea. The creation process takes place at the speed of light so therefore nothing can move faster than the speed of light and still maintain its essence. Everything we see, everything we do is all a realization of the idea. We are the product of the supreme thought. The universe is the perfect form of art and is constantly in flux.

If you look at a specific work of art, and if you really knew the artist well, you would immediately be able to tell that this specific piece was the work of that artist, even if you were unfamiliar with the piece itself. That is because you can see the hand of the artist at work. The painting itself reflects the mind and will of the artist. It should therefore follow that if God created the universe, we should see the hand of God at work in every little piece of the universe, no matter what God is, if the universe was created from a specific thought or idea, we should see evidence of that idea reflected in the art itself. But, we don't know the artist, we don't know God. In order to understand or know God we would have to assume that everything we see in the universe is a reflection of God. This is exactly the predicament we find ourselves in.

We question the existence of God and ask where God is? But everything around us must be the essence of God, no matter what God is, no matter what the creator is, we must be a product of the creator. If the universe was created from pure gravity, then we must be the essence of gravity. No matter what you call it, the force that created us is reflected in us, it is a part of us.

But why would God create such a universe, why not create a universe in which there is no suffering or death?

CHAPTER 13

Perfection

The Universe is the way it is because it is perfect. As Gottfried Leibniz said "God created the best of all possible worlds". What possible evidence do I have of this? How can I sit here and say that this violent world in which babies get beaten to death by their own mothers and fathers could possibly be perfect?

I wish you would read my previous books. Especially the first one in which I ask this very question, if you look at all the injustice around you, how could any of this have purpose? I'll tell you how, because despite everything I have seen and everything I have been through, I still love life, I love this world, it is so bitter-sweet and beautiful that sometimes I want to cry and laugh at the same time. Why is this? If things are so terrible, then why do we care? Why do we fight to make things better? Why do we love? If life is really so random and pointless, why do we bother? The answer is because everything is perfect. The universe makes us care about life.

People love life, sure we moan and complain about everything and we want things to be easy, but we love life. We fight to protect it, we do everything to preserve it, we only moan about the negative because we know how precious life is, we want everyone to enjoy it, we want it to be PERFECT, we want to fill ourselves up with the pure joy of it.

This is precisely because everything is perfect. A life in which you had everything handed to you would be pointless and dull. You would have no motivation or appreciation for anything. If people didn't die there would be no such thing as love. Why would you care for anyone or truly love them if they will be here all the time anyway. And, as much as it pains me to say this, if my life was any easier I would not have bothered to write this book, if I had everything I wanted I would not have had the motivation to wonder about the world around me or care to do anything to change it. There would always be tomorrow. Life would be meaningless once you have everything.

But the universe is not only perfect because I am deluded and have convinced myself that it is so in order to justify not slitting my own wrists from sheer misery. I believe that if you look at the physical perfection of the universe you will see the same thing.

Aside from being a perfect idea I think the universe is actually a perfect construct too. Going back to the artist Giotto, who was asked to prove his value as an artist to the pope, Giotto then draws a perfect circle. Two things are happening here, firstly, every little thing in the universe happens for a reason, the one reason this story becomes so popular and endures the test of time is because it is a reflection of the true nature of the universe, the second is that Giotto is literally saying that true skill or perfection can be represented by a perfect circle.

I think the universe itself is a perfect circle. This is a very difficult statement to make because of the elusive or dualistic nature of the universe. On the one hand the universe is only an idea or an event, but to us stuck inside the universe as part of the universe it appears to us as a physical construct.

The reason for this is simple. We can either choose to think of the universe as a place in which we are living and moving through time. In this case we cannot question the universe; we must simply live as natural animals or creatures purely trying to survive.

On the other hand, we can ask the big questions like is there a God and what is the universe? But if we do this we must realize that we are only a part of the universe we are not separate from it. In this case we must hypothetically step outside the universe and question creation as a whole; we cannot sit inside the universe and answer the question of the universe. A good analogy of this would be a person trying to answer how the human heart works by using his/her own heart; there is no way a person can look at their own body and answer this question in its entirety. In order to truly understand how the human heart pumps blood we would need to have a separate body to dissect and probe. Even if you use an MRI of your own body you still need this external reference. The same holds true for the universe, we are the universe, we are a part of space and time, in order to answer the entire question regarding the universe we would need another copy of it to study, but seeing as how we don't have one, we could hypothetically remove ourselves from this one and try to answer the riddle. If you do this you must bear in mind that you are questioning the universe as a whole.

And what is a whole universe? A whole universe would be all of space and time, from the start of time to the end of time. From inside the universe we cannot see it as a whole; we can only look backwards in time, no matter what direction we look in from inside the universe we are only looking backwards in time. So what would the universe look like if we had one in front of us? Again, this is not a question I am comfortable answering because of the perfect nature of the universe which I will explain fully.

Assuming we could see the entire universe from an external position I think the universe would look like a perfect two-dimensional flat disk or perfect circle or sphere. In fact I think in time this is how we will see the universe from the inside too. Please, I cannot emphasize enough that the universe is NOT a place therefore it has no OUTSIDE. The universe is an event. We can only see it from our perspective inside and as part of the unfolding event. From our point of view it will always appear spherical. But the universe never really comes to permanent rest in this or any other form, when we look at the universe we only see a single moment in time. The universe itself is constantly evolving and moving.

We are looking for a way of defining or creating a mental picture of how the universe works.

In this regard, think of the universe as a large blank rewritable DVD or data storage disk. At the dawn of time God starts to write the story of the universe onto the blank disk using the speed of light, all of the past, present and future is written onto the disk and can be read by an observer exactly like playing a movie, each frame of your life is stored on this disk and is eternal. The disk itself is pure energy when you write on it the energy is converted to mass. It has two states.

But because it is *perfect* the universe doesn't stay in this state. If you think of a perfect circle it has no start or end, you can run your finger or mind along the circumference of a perfect circle and never know where it starts or ends. So the universe if it is to be perfect must also have no start of time or end of time. Once our blank disk is full of data we find that it starts to erase itself, all the information gets stripped from the physical universe, the story of the universe goes back to being only an idea and the physical matter or disk is wiped clean. Matter is converted back to energy. And now a new story can be written. This is simply an analogy of what is happening during creation and in what we have come to think of as 'black holes'.

In creation we find that God created the universe using pure energy. For every action there has to be an equal and opposite reaction, this is basic physics, in the creation event when this universe is created an equal and opposite universe or anti-universe is created. I think the religious interpretation of this process would classify these two worlds as heaven and hell. But we must not think of this anti-universe as a separate place, it is a part of this universe. If we have matter we have anti-matter. In our disk analogy you can simply think of it as the reverse side of the disk.

If you remember that the universe is not actually a place or even a disk but rather it is an event or something that is happening then remember also that there are always two sides to a story. Positive and negative. Start of time and end of time.

Dark energy or dark matter is simply the amount of energy or matter left in the entire overall process. So if the disk is only one quarter full, dark energy and dark matter are the remaining three quarters. They are the part of the event that hasn't happened yet. The entire universe is a closed system because the amount of energy is fixed. Exactly like a rewritable DVD that can only store a fixed amount of data. You can wipe it clean and save new data on it over and over again. But it does not have infinite space. There is not an infinite amount of energy or matter in the universe. There are only infinite cycles. The universe is not something that happens and stops, the universe recycles for all eternity.

Because we are made from the essence of God, we are made from this same energy that keeps recycling, some part of what is really happening is ingrained into our very being, instinctively we know that there are two worlds, religions call them heaven and hell, science would call them matter and anti-matter, on our physical circular disk we would simply see two reversible sides.

Earlier, in the chapter concerning dreams I reserved the right to make a final comment on the nature of dreams and the occult. I think that maybe because we are made from this same energy that recycles into infinity, maybe some people are able to tap into this energy and can therefore predict the future. Since the energy itself is God, God survives from one cycle to the next. God knows no time. God knows all possible realties, all possible outcomes. Since we are made from this energy maybe some small part of this knowledge is saved inside all of us.

As time goes on the universe as we know it is converted back into pure energy by 'black holes' which is simply mass that has reached the speed of light, once it is totally converted the formless energy is converted back into the universe, and so the cycle continues into infinity. Now nobody could ever tell where the universe started and where it ends, there is only this continuous flow of energy into matter and matter back into energy. Did God create the universe or did the universe create God? What came first the chicken or the egg?

Man will never be able to answer this question.

This is the perfect form of art, the artist becomes the art because he/she wants to transform themselves, but what is the ultimate purpose of the art? To reflect or be the essence of the artist. For anyone stuck inside the universe, they would suffer under the illusion of time and space, intelligent beings would look back in time and see a single point in time from which the universe starts, and they would see that the universe expands and will have to end somewhere in the future.

If anyone could somehow exist out of time and space they would see a perfect work of art, which is totally impossible because there is no such thing as outside the universe, there is no starting point, there is no finish line. You cannot get outside the universe. The end of the universe would be the end of time. And since time cycles into infinity you cannot get to the end of the universe.

The universe only exists to entities inside the universe trapped under the illusion of time and space. We are the idea or the story. Outside of the universe the story has no form. I would further say that there is no such thing as multi-verses or multiple universes, because there is nothing outside of the universe. There are no black holes leading out of the universe either. There is only the universe and God (Idea of the Universe).

So why are some people so convinced that there are alternate universes or multi-verses (Heaven and Hell)? Hold on, we are getting to that.

CHAPTER 14

God

We are made in the image of God, so we are in fact God. If everything in the universe is made from the same primal everlasting energy then we are also made from the same energy. If you accept that E=MC^2 is right then you must accept that the brain inside your head along with the rest of you is made out of energy. And if this energy can create your advanced intelligent and complex brain, imagine the overall power and intelligence of all the energy in the entire universe combined.

Christianity talks about the son of God being Jesus. Jesus message to people is that there is no sin. He tells people that he has died for our sins. Well, all that is true, because in fact we are all the children of God, we are made from the essence of God. There can be no such thing as sin because we are God, everything we are is what God wanted us to be, every action we take is exactly what we were meant to do. God did die for us, because God transformed himself / herself / itself into the universe. The energy became the mass. God is the energy.

This is all part of the perfection of the universe, we have free will, but the universe is so perfect that no matter what we do, that is exactly what we were meant to do because there is no right or wrong.

For anyone asking where God is? Why does he allow such suffering? I would reiterate that we are God, we are free to create our own reality, and we build society. If there is suffering in the world it is all due to our own doing. I said 'IF' there is suffering because I am no longer convinced that suffering exists on its own. If it does exist it is something we all share. I don't believe that anyone inside the universe is any better or worse off than anyone else. Am I crazy? Have I not seen footage from Ethiopia showing those starving children? This is an argument that goes down to the very nature of human beings, are those people only suffering? Do they not know any other form of reality? Do they only suffer? Well, then why are they still alive? Surely they must feel something other than suffering; they must feel something that makes them want to stay alive. I believe that we all have that same something in equal measures. Call it hope or faith. We all suffer and yet we all want to stay alive. This is how incredible this universe is, this is how incredible God is, life is perfect because it is so fragile and precious.

So what about the starving children? Is that their fate? No, we have to change the world, we have free will. We can be greater than this, we are God, we can do whatever we want inside this Universe, this is our story; I am not cold and heartless to say that these people are not suffering, but we have to see the truth first in order to change the world. We must understand that the suffering we see is our own doing. If we blame God, we are blaming ourselves.

Do not forget that time is relative to the observer, the idea that we are in fact moving through time is only an illusion, you will always be here in this moment in time, so you must speak up and create the world you want to live in because you are here for eternity. Remember the artist who always keeps dabbing at his painting, always wanting to change something. It is the same with God. God has given us free will; we can create any universe we want. Nothing is predetermined. How can this be if the past, present and future already exist?

This is part of the perfection of creation. Remember those alternate universes or multi-universes, well, if this universe recycles into infinity, then every time it recycles we will have a chance to do things differently, this is why we have free will. We can create an infinite number of universes or alternate realities, but we can only exist in one of them at a time, time will always be relative to us.

Remember also that on an infinite scale, although you can create an infinite number of alternate realities, you can also create an infinite number of this reality. So although we seem to age and move, although it seems like time marches on and planets and stars are constantly being born and destroyed, we cannot be destroyed because we will always exist in this exact time and space thinking this was the present.

We are a recorded story. The physical world is simply the material on which the story is written, but once that material is removed, the story has already been told, and will last forever. There is no real distinction between what we really are. Once again scripture speaks of mortal man and spiritual man, this is the dual nature of our being, we are both physical as represented in the world around us and we are spiritual in the sense that we are only an idea.

This universe is heaven and it is eternal. This is Gods promise to us, a place in which we can exist for all eternity without the fear of death. We will always be with our loved ones. But remember there are always two sides to the story. You can choose to be happy or you can choose to be ungrateful and miserable. Positive or negative? All you need is faith.

God is all around us and part of us. If you still think that I must be crazy or if you need more proof that what I am saying has any morsel of credibility I can give you more. If indeed I was right and we are all partly a recording on a super massive and infinitely complex 3 dimensional disk than surely there must be some physical evidence. Well, I think there is. Bear in mind my physics knowledge is very limited.

If you look at any recorded footage, like a movie, that footage is always recorded in frames, now, if you slow that footage down enough you will actually be able to see the missing frames, depending on the quality of the recording you might have a film that runs very jittery because there are too few frames or you might have a movie that appears very real and fluid all depending on the amount of frames.

If I am right and the physical universe is a recorded version of our story then surely if we zoom in really closely, I mean really zoom in until we can see the finest particles or sub-atomic building blocks inside the atoms that make up the universe, on this microscopic level we should find those missing frames. I think we have, if we view these tiny particles we see that they simply jump from one position to the next.

Researchers in the field of Quantum physics are talking about "the Planck scale" named after Max Planck. In terms of size, the Planck scale is unimaginably small (many orders of magnitude smaller than a proton). They believe that these tiny particles disappear down a black hole and emerge a fraction of time later in the future and that at this level the universe is a seething network of black holes. I disagree. I think they are simply the missing frames. This universe is of such high quality and on such a grand scale that we really believe that we are REAL. We think we are separate from the universe and that we come into the universe when we are born and we leave it once we die. The truth is, this physical realm is simply the recorded version or manifestation of our story. We do not physically exist apart from the universe. We cannot leave the universe because we are not separate from the universe.

This theory about us being a recorded story or expression of the idea also explains the "Twin paradox" that I mentioned much earlier which many consider to be a flaw in the theory of special relativity. For anyone unfamiliar with this paradox, it simply asks why is it that if you take twins, place one on a spaceship and speed him up to close to the speed of light and then slow him down to join the normal flow of events, why has this twin aged much slower than the twin who was moving with the normal flow of events, because if special relativity is correct then to each twin the other one was the one that was travelling or speeding, so why is it that one ages much slower than the other? I think it's simple; it's like hitting fast forward on your DVD recorder and flying over a larger amount of frames in the same amount of time. You will have aged normally but the time in the story has advanced considerably. Therefore all of time must in fact already exist; there must be a fixed amount of time to the universe.

Most physicists would not agree with me, they think that time is in fact being created as the universe expands. This is only true for us as part of the universe experiencing time as it unfolds. But there must be an overall universal time. The universe is in fact counting down and time is running out as we move along. I theorize that the universe is a 60 billion light year clock, this is the period it takes to complete one full cycle, our cycle started about 13.7 billion years ago, maybe even closer to 15 billion years.

I have posted a visual representation of what I think is happening over time on Youtube. Please feel free to visit this site and have a look. Bear in mind that this is only an attempt to explain what is happening over time, it does not imply that the universe is a 3d torus. The universe is only something that happens in endless cycles.

http://www.youtube.com/watch?v=I53sQZvKLFw

Think of this video as a visual representation of $E=MC^2$. Approximately 13.7 billion years ago energy converted into mass. This conversion happened at the speed of light. As energy moves outward in all 360 degrees following the vertical circles it stretches the universe (fabric of space-time) and causes it to expand at twice the speed of light in all directions (horizontal circles).

This is why the universe appears to expand faster than the speed of light. It is not because mass or space can move faster than light or that it ever could (as predicted by the theory of inflation which is used to explain the big bang). It is simply that space is expanding in all directions at the speed of light.

This is also why every object inside the universe will see itself as being at the centre of the universe. It is not because the universe is a place and we exist in the centre. The universe is an event, if any intelligent species anywhere in the universe looks back at when the event started they will see that it started roughly 13.7 billion years ago and it is expanding at the speed of light in all directions. Space itself will therefore be growing at twice the speed of light.

If you look at a super high quality photo with millions of pixels, it will look like a solid or fluid image, but if you zoom in really close, you will eventually see the little individual dots or pixels that make up the image. Couple this with the theory that we are moving through time is just an illusion and that we in fact exist in this one moment for all eternity, now, when you add movement, like the spin or cycling of the universe at the speed of light, it will appear as if the dots / pixels are animated.

This is what time is and why time varies at different points in space or at different velocities. Time is an illusion created by the movement of energy. The entire universe must be recycling at the speed of light.

We exist in two forms, the physical world and the spiritual world. But there is no real difference between the two because the transformation from thought to form, from artist to art is so perfect that it does not matter. And why? Because creation and the creator are so perfect.

What happens to us when we die? Well, we don't really die; we are always here in this time and space, always clinging to life, always wanting more of it, always feeling so filled with sorrow and joy that we simply want to burst. Death is nothingness. Once we die our physical bodies are converted back into energy. Eventually all the matter in the universe will be split back into energy and information by the 'black-holes' which exist at the centre of every galaxy.

CHAPTER 15

Heaven

This universe is heaven. Really think about the concept of heaven. Religious people will tell you that heaven is an ideal place of great beauty in which you reside with God and are reunited with all your loved ones to live happily ever after.

That is exactly what this universe is. There is nothing outside of the universe. The universe has no outside; it has no start and no end. It is a perfect circle. The universe is God and God is the universe. This is the perfect work of art. You will always exist in your own specific time and space. This is your slice of heaven. You will always be with your loved ones. There is no place more beautiful than the universe. Surely you must see how greedy it is to ask for more than this. Look at nature. What could possibly top all of this? God did make the universe for a reason, the reason being that it is the most beautiful work of art there ever will be. There is no separate heaven. All you have to do to realize that what I am saying is true is open up your mind. This is why we have the ability to think, our thoughts will lead us to the truth. Do not close your mind, read some books and keep learning about the world, that way you will see the perfection in everything.

The best part is, nothing is rigid, don't sit there thinking your life is so horrible you don't want to be stuck in it for all eternity. You have free will. You always had. Even the little child version of you is still there in the past right now deciding on two different options, which one will he /she pick? That decision will change his /her whole life. You exist at every point and time in your life, and at every point in time you have free will, you can change your life to be whatever you want it to be. Do not let anyone else control you, you have every right that anyone else has, you are God. There is no wrong or right. I am not afraid to tell people this. I am not worried that if people believe me they will suddenly go out and start killing and pillaging to get everything they want. I know this because that is not the nature of God; I know this because you don't create something as wonderful and as beautiful as the universe if you are destructive by nature. I trust that if people do believe me, they will see the beauty of SELF and the meaning in everything we do. I have FAITH in mankind. I have faith in this God, because this God is not petty, or selective, this God is everything, this God is all encompassing. This God will unite us.

This is the message. God is everything. You don't have to pray or worship idols; you don't have to be subservient or pious. Be free. God is everything. You are a part of God.

Seriously think about everything you have read in this book. The answers are all around you. Go back and read through your scriptures and bibles, and try to apply everything I have said about the nature of God and life to what you read, you will see that I am right, or very close to being right. I do not claim to know everything about creation, I simply state what I feel very deeply because I know that some part of it must be true. Test me. Apply this philosophy to what you see in the world.

Every person you meet, every plant, every animal every star you see out in the cosmos, is a part of you, we are God, and we are the universe. We created this world because we wanted to experience all those things. You must embrace this thought, there is no sin, there is no wrong, there is no right, there only IS.

This does not mean you must passively or apathetically accept everything the way it is, you have free will, you can change your world, you must embrace the world for what it is, this will give you the strength you need to change the world to what you want it to be. I know that deep down inside we all want the same thing.

I said the universe and our physical self is simply a recorded version of the idea, this does not mean the record is cast in stone; it is simply the physical form in which the idea is manifested. That is further proof of how perfect everything is, this is a living recording that can change according to our free will. It is empowering to know that everything you do is meant to be.

You think this is impossible. How can a recording be in flux? Have you seen those Blu-Ray movies or DVD's in which you can select different endings for certain movies? Even some computer games have different endings now. We do this using our limited technology. Imagine if you had access to all the energy in the universe and infinite time what you could create.

Many religions talk about heaven and hell and even physicists are starting to talk about alternate universes. There is no other universe, there is only this one, but you can change your universe at any point in space and time, so in a manner of speaking, they are right, there are multiple versions of this universe, but you will only ever experience the universe you create. Don't think that you can leave this universe through a wormhole and get a better life; there is only the universe you create. You see, even Nietzsche was right with his philosophy of perspectivism.

And what about death? Death is the ultimate challenge. You must conquer your fear of death to be happy. Most people try to conquer death by thinking they can beat it, or outsmart it. People are drawn to the church because the church promises life after death. Science is drawn to alternate universes thinking they can create wormholes and escape time. But the only way to conquer death is to embrace it. This is not a defeatist attitude; I am not saying you must embrace death because you can't beat it. I am saying the only way to beat death is to embrace it.

Death is a gift. Once again, people's fear of death comes about from a lack of being able to follow a single thought to its logical conclusion. You have to sit and think about living forever, if you can really contemplate this concept, sooner or later you will start to feel claustrophobic, as beautiful as the universe is, as big as it is, if you had to live forever the universe will become a coffin, you will be buried alive for all eternity with no hope of escape. Death is your escape. Death makes life precious and beautiful.

CHAPTER 16

The Circle

I must continue to emphasize that we are created in the image of God and therefore everything we do reflects the true nature of God and of the universe. There is hidden meaning in everything. There is even hidden meaning in people who say that life is totally random, everything happens for a reason.

I believe that there are clues all around us, the fact that the wheel is mans greatest invention is not simple coincidence. The circle is Gods greatest invention. Why did the introduction of the zero alter all of mathematics? The circle is the key. The signs are everywhere, of course anyone could read anything they like into almost any aspect of life and call it destiny. This is part of the perfection of life. We will never know the truth, science can go chasing after the stars all they want but at the end of the day, it will all come down to what you believe, not because it will make any difference, not because you will go to heaven if you believe in Jesus, it will all come down to what you believe because that is all that will get you through life.

You can live your life thinking there is no magic and no reason or you can find magic and reason in everything. Faith is the path to heaven. Let me give you a perfect example of finding meaning in everything. I have faith that my name means light for a reason, even though my mother had no idea why she named me. I have faith that my surname is Khan which means leader for a reason, even though my father and his father simply inherited this surname. It could all be irrelevant and random, or you can see the magic in it. I believe that it is my destiny to lead people to the light, it's in my name.

I will stand by my view that the pursuit of scientific knowledge and understanding is the highest calling because this will lead us to the eventual conclusion that life is not a matter of physics, one day, be it a billion years from now, we will transfer our thoughts into a computer and we will see that every person is essentially the same, we all have the same thoughts, emotions and feelings, only life with all its beauty and suffering creates individuality.

Are religious people foolish in their faith? No. Only misguided maybe. I have always harbored a great contempt for religion, not because of their message but because of the politics behind it, religion should not endeavor to segregate people. There is no single group of people who can claim to be more favored by God. God is in everyone. You do not have to be pious in order to receive God's blessings. Real faith should serve to comfort and aid people on their journey through life, it is totally acceptable to have faith in the notion that you are here for a purpose and that goodness is the path to happiness. The bible will tell you that if you have Faith in God you can do anything. Life will teach you that if you have faith in yourself you can do anything. That is because you are God.

Coming back to circles, why are circles so important? Once again God has shown his true mastery and SUPREME INTELLIGENCE here; circles have a very peculiar aspect about them. No matter where you are in the universe if you take a perfect circle of any size, measure its circumference and then divide that figure by the diameter of the circle you will get PI (π). Pi is usually rounded off to 3.1416. No matter what size the circle is you will always get this result. And what's so fascinating about this result? Well, like I said, Pi is usually rounded off to 3.1416 but in actual fact, Pi is infinite.

So far they have calculated up to 5 trillion decimals of pi and still there seems to be no end in sight. What's more is, there seems to be no sequence or repetition to the decimals either, here are the first few decimals of pi:

PI=3.14159265358979323846264338327950288419716
939937510....

As you can see, pi simply goes on and on, seemingly into infinity. How can a perfectly closed shape give you such a result? No matter where you are in the universe a perfect circle will always give you pi. I think this is the ultimate clue to the true nature of the universe. It is a perfect circle and it is infinite or eternal, further more it is never rigid but always changing.

Look around you; do you have a clock on your wall? What is a clock? Time moving around in a circle, why is that the way we represent time? Is that the only way to count out time? Certainly not, we live in a digital age where we have digital watches, so why is a circular clock still so popular, it's because we are literally looking at the circle of time.

Look out into the cosmos and what do you see, planets revolving around stars, moons revolving around planets, bored farmers making crop circles, rainbows (semi-circle of light), everything making circles. The art is a reflection of the artist. The answers to the universe are in the universe. You will never find perfection inside any single object or idea inside the universe, but everything inside the universe is a part of the perfection that is creation.

CHAPTER 17

Technology

The answers are literally everywhere and nowhere more so than in humans themselves, look at our technology, look at our art what are we doing? Every day we move closer and closer to creating a virtual world with startling clarity and reality, we create games in which the characters have free will but the final outcome is always predetermined. You can run around inside a virtual world forever and yet it has a fixed size and can fit onto a circular disk. We use light in the form of lasers to read the information on these disks.

Scientists are talking about creating nanobots that will one day be able to transfer all our thoughts into a computer or maybe even simply being able to download all our thoughts straight from our brains. Assume this will be possible in the future, even if it is a billion years from now, what will they find once they put all our thoughts onto a computer? You know what they will find, they will discover that every person is essentially the same, ultimately we are a single thought, do you really think that you are so different from everyone else? The only thing that makes us unique is our human bodies and our circumstances and experiences. The only thing that makes life so wonderful is the fact that one day we will die. We only have a limited time and space. Although we will exist for all eternity in our own time and space, we will not live for all the billions of years of time and space.

So what will happen when they finally download our thoughts into a computer? The first thing they will find is that there is only the single thought or mind. And what joy could there be in being a computer that will never die or ever experience any external sensations, or sense of self. How lonely will that superior mind be? When that time comes we will miss our human bodies, we will yearn to eat a peach or make love or feel the wind blowing through our hair, we will no longer want to be a single thought, we will long for individual experiences. Don't you see it? We are a reflection of God. Everything we do tells us everything about creation.

A billion years from now or even 10 billion years after that, those machines will be so smart that they will be able to program an entire universe exactly as we see it now, right down to the last grain of sand, we already have 3d software packages that can mimic gravity, simulate smoke and dust particles down to the last grain, they can even simulate water and fire. And it all looks so real when it is in motion, all these digitized pixels coming together to create the illusion of life. A billion years from now these super-computers could probably even program human beings into the simulation and make them think that they are alive. They will store all this data on some kind of storage disk. We are already talking about creating computers made entirely out of light because light can carry information.

Can you see what perfection is? Those people that they program into their simulation of the universe could follow the exact same sequence as we did, because they will really only be a reflection of the thoughts of their programmers. Where will it end? Each universe will get smaller and smaller, but because size is relative, the worlds that each generation creates will seem just as vast and splendid as our own universe.

There is no start or end to the universe. The universe has no time, space or size. The universe simply IS. Nothing is ever cast in stone, you can choose to believe this or not. Don't worry about it, there is no right or wrong. You have free will. You will always be right here, right now, maybe next time you will believe and your life will take a different path.

Beautiful perfection.

There is no END.

ACKNOWLEDGEMENTS

Special thanks to Jon Swinbourne and Ted Brandes two guys I met online who really made me think about what I was saying and challenged me on almost everything I theorized. And even though I may not have convinced them that I am right, I must acknowledge that they have helped me understand more about the physics of the universe and also helped me to formulate my thoughts.

It is based on these discussions that I have amended my earlier release of this book to incorporate the knowledge they have bestowed upon me.

Thank you as always to my wife Bronwyn for supporting me through all my ups and downs.

Other books published by Author include:

Terms and Conditions. First Published in 2009

ISBN NO. 1449911027

Eternal Life, The Universe and Happiness. First Published in 2010

ISBN NO. 1453792465